Investigate the Possibilities

Elementary Astronomy

THE UNIVERSE
From Comets to Constellations

Student Journal

Tom DeRosa
Carolyn Reeves

THE UNIVERSE

From Comets to Constellations

Student Journal

First printing: April 2014

Master Books® is a division of the New Leaf Publishing Group, Inc.

ISBN: 978-0-89051-798-7

Cover by Diana Bogardus

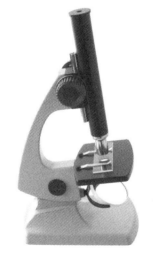

Unless otherwise noted, Scripture quotations are from the New International Version of the Bible.

Please consider requesting that a copy of this volume be purchased by your local library system.

Printed in the United States of America

Please visit our website for other great titles:
www.masterbooks.net

For information regarding author interviews, please contact the publicity department at (870) 438-5288

Master
Books®
A Division of New Leaf Publishing Group
www.masterbooks.net

Table of Contents

Note to the Student

Record your ideas, questions, observations, and answers in the student book. Begin with "Think about This." After you read "Think about This," try to recall and note any experiences you have had related to the topic, or make notes of what you would like to learn.

Record all observations and data obtained from each activity.

You should do at least one "Dig Deeper" project each week, recording the projects you choose to do, along with the completion date in a notebook or journal. Your teacher will tell you how many projects you are required to do. The reason for the large number of projects is to give you choices. This allows you to dig deeper into those areas you are most interested in pursuing. Most of these projects will need to be turned in separately from the Student Journal.

Record the answers to "What Did You Learn."

The Stumper's Corner is your time to ask the questions. Write two short-answer questions related to each lesson that are hard enough to stump someone. Write your questions along with the correct answer or write two questions that you don't know and would like to know more about.

Some of these experiments should be done with the help of adult supervision. They have been specifically designed for educational purposes, with materials that are readily available.

Investigation #1
What Is the Universe?

Date: 9/19/23

The Activity: Procedure and Observations

Follow the directions to make a model of the distances of the planets from the sun.

1. Each penny represents a distance of 1 AU (astronomical unit). Use the chart in your text to find the approximate distance of each planet from the sun. Write the names of the planets and tell how many miles or km each planet is from the sun.

Planet	Distance to the sun
Mercury	.4 AU
Venus	0.7 Au
Earth	1.0 Au
Mars	1.5 Au
Jupiter	5 Au
Saturn	9 Au
Uranus	19 Au
Neptune	30 Au
Pluto	40 Au

2. Predict how many pennies it would take to represent the distance to the nearest star from our solar system. __100__

Stumper's Corner

1. _____

2. _____

1. Why do scientists sometimes use models to explain things in nature?

Cause it helps other People to understand

2. What is one problem with using the penny model to help explain the solar system? because space is so big that we would Need the hole ocean

3. What kinds of things are found in our solar system? Rocky Planets, Planets, Stars.

4. Name the four rocky planets in order. Mercury, Venus, Earth, Mars

5. Name the four outer gas planets in order. Jupiter Saturn, uranus, and Neptun

6. Explain what is meant by an AU unit. is a astronomical unit

7. Where are the solar system's asteroid belt and the TNO region?

8. What is a light-year? Why are AUs not used to measure some distances between objects in the universe? becaus All units are to small a lightyear is the diodised light travel in one year

9. Briefly explain each of the following: solar system, galaxy, cluster of galaxies, and universe.

10. "Milky Way" and "Local Group" are the names of two things found in space that contain the earth. What is each?

11. What is the nearest star to our sun? Is it in our solar system? Is it in our galaxy? Yes

12. Briefly describe one method scientists use to estimate the distance between objects in space they Mesure from one angle of eart than in six Moths the other angle

Investigation #2
Spreading Out the Heavens

Date:

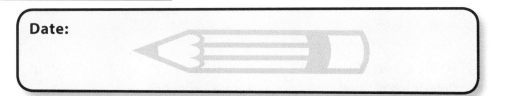

The Activity: Procedure and Observations

Part A.

Partially blow up a large balloon and put colored dots all over the outside. Space the dots somewhat evenly apart from each other. Continue to blow into the balloon so that is gets larger. Notice that the dots get farther and farther apart as the balloon gets bigger. Now gradually let the air out of the balloon and notice that the balloon gets smaller and the dots get closer together. Blow up the balloon again and observe. Write about what you observe.

_____

_____

_____

_____

_____

_____

_____

Part B. Record times

1. How many marshmallows did B catch in 10 seconds when no one was moving?"_____

2. How many marshmallows did B catch in 10 seconds when A was walking slowly toward B? _____

3. How many marshmallows did B catch in 10 seconds when A was walking slowly away from B? _____

Repeat this activity two more times. In which situation was it easier to catch the marshmallows? _____

Stumper's Corner

1. _____

_____

2. _____

_____

1. What scientist discovered that there were other galaxies in the universe in addition to the galaxy our earth is in? _____

   Edwin Hubble

2. What evidence did Edwin Hubble discover that caused him to conclude that galaxies are moving and getting farther away from the earth? *A Red shift*

3. Before the time of Hubble, did scientists believe all the stars in the universe were in the same galaxy? *yes*

4. Which color in the visible spectrum has the longest wavelength? ___

   Red

5. Is the bluish/violet end of the visible spectrum made up of shorter waves or longer waves? *shorter*

6. What major shift in thinking about the solar system came from scientists like Copernicus, Galileo, and Kepler? *H Made People beleive there was a Sun scenerd solar sistam*

7. The "big-bang" theory is based on what two pieces of evidence? Does this prove that the big bang actually happened? *No*

8. Briefly tell about the "nebula theory." Does it attempt to explain the origin of all the stars, planets, moons, comets, rocks, and dust in the universe? *No it does Not explain*

9. What instrument was available for Galileo, Kepler, and Hubble to use that Copernicus did not have? *a teloscop*

10. Give the shape of two different kinds of galaxies. _____

11. All galaxies appear to be moving. Why are we unable to look at them and tell that they are moving? *becoos there to far away*

The Strange Behavior of Space and Light

Date: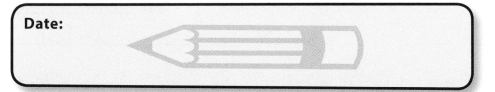

The Activity: Procedure and Observations

1. Put a string around a sphere (a ball) at its widest point. Measure this length of string with a metric ruler. Record this distance in centimeters. _____

2. Measure the length and width of a flat piece of paper using a metric ruler. Record this distance in centimeters. _____

3. Cut a strip of white paper 4 or 5 cm wide and about 28 cm long. Without twisting it, join the ends and tape the ends together. Use a colored marker to draw a line through the middle of the strip of paper. Don't let your marker leave the surface as you draw. Keep drawing until you end up where you started.

4. Cut another strip of white paper 4 or 5 cm wide and about 28 cm long. Twist it over, join the ends, and tape the ends together. Use a colored marker to draw a line through the middle of the strip of paper. Don't let your marker leave the surface as you draw. Keep drawing until you end up where you started. What do you observe about the line you drew? How is it different from the previous line you drew? _____

_____

_____

_____

✏ Stumper's Corner

1. _____

_____

2. _____

_____

1. Why is a Mobius strip difficult to measure? _because_ _it only has one serface_

2. According to Einstein's special theory of relativity, what is the fastest speed that anything can reach? _light year_

3. There were six hypothetical explanations for how light was able to reach across the universe in less than millions of years. Briefly explain one of these explanations. _Light could reach_ _earth verry quickly from_ _any where in space_

4. What is one way in which Einstein's theories of relativity add to Newton's laws of motion? _____

5. Are Newton's laws of motion always correct? Under what conditions might they not be correct? _____

6. Does Einstein believe that light always travels in a straight line when it passes through outer space? _no_

7. Einstein studied time and what four other factors as part of his theories of relativity? _____

8. What did Einstein name his two theories of relativity? _____

9. What is the basic difference between Einstein's two theories of relativity? _____

Investigation #4
Kepler's Clockwise Universe

Date:

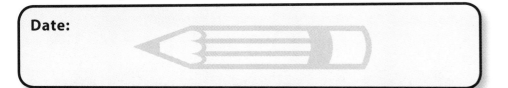

The Activity:
Procedure and Observations

Draw a circle around a thumbtack on a sheet of paper as instructed. Label the thumbtack as the "sun."

Follow the instructions to draw a variety of ellipses. Some of the ellipses you drew on another sheet of paper will be too large to draw in your journal, but try to draw a small ellipse below. Label the two foci. Describe some of the shapes of the ellipses you drew.

Drawing Board:

Sun

1. _____
   _____

2. _____
   _____

? What Did You Learn

1. Before the time of Copernicus, scientists believed that the sun, the stars, the moon, and the planets orbited the earth. Explain what Copernicus proposed that disagreed with this. _____
   _____
   _____

2. Copernicus believed that the planets orbited the sun in perfect circles. What kind of orbit did Kepler propose that planets followed?
   _____
   _____

3. What is an elliptical orbit? _____
   _____
   _____
   _____

4. The earth gets a little closer to the sun at one phase of its orbit. Does the earth get warmer when it is closer to the sun? _____

5. What causes the different seasons of the earth? _____
   _____
   _____
   _____

6. What is the difference in the orbits of the inner planets and the outer planets? _____
   _____
   _____

7. Do comets orbit the sun? What kind of orbits do they follow? _____
   _____
   _____

8. Kepler discovered three basic laws about planetary motion in the 1600s. Do modern astronomers still use these same laws as they study the planets? _____
   _____
   _____

9. Did most scientists immediately agree with Copernicus when he proposed that the planets revolve in orbits around the sun? _____
   _____
   _____

☐ Dig Deeper completed in notebook.

Date:

The Activity:
Procedure and Observations

Part A.

Drop a flat sheet of paper and a marble at the same time to see which one hits the floor first.

1. Describe what you observed. _____

_____

_____

Crumple the paper into a tight ball and drop it and the flat sheet of paper at the same time.

2. Which one hit the floor first? _____

Now place the other objects near the edge of a flat table and quickly push them off the table at the same time with a meter stick or flat board. (Be careful not to use objects that might break or that might damage the floor.) Other students should stand where they can have a good view of when these objects hit the ground.

3. Which object hit the ground first? Repeat this several times until you can determine which objects hit the floor first. _____

If you can find a shelf that is higher than the table, repeat this activity and record your results. (Use a pad if necessary to prevent damaging the floor.)

4. Which objects hit the floor first this time? _____

_____

If possible, view the NASA video showing an astronaut dropping a hammer and a feather on the moon at the same time.

5. If you are able to see the video, try to think of an explanation for what happens.

Part B.

Use an empty paper towel roll. Cut it lengthwise into two pieces. Prop one of the pieces up so that one end is about 15 cm high and the other end is on level smooth flooring. Release a marble from the top of your ramp and observe it until it stops rolling.

Measure the distance the marble rolls across the floor with a metric ruler. Reposition your ramp so that one end is 10 cm high and the other end is on level, smooth flooring. Release the marble from the top of the ramp and measure how far it rolls across the floor this time. Reposition your ramp once more so that one end is 5 cm high. Release the marble as you did before and measure the distance it rolls across the floor.

Height of ramp	Distance marble rolls on the floor
15 cm	
10 cm	
5 cm	

1. _____

   _____

2. _____

   _____

❓ What Did You Learn

1. Why did a flat sheet of paper fall more slowly that an equal size wadded-up sheet of paper? _____

   _____

   _____

   _____

2. When air friction isn't a major factor on a falling object, do all objects dropped at the same time, and from the same height, hit the ground at the same time?_____

3. An astronaut dropped a feather and a hammer on the moon at the same time from the same height. Why did they hit the ground at the same time? _____

4. Around what time period did scientists begin to use telescopes to study planets, moons, and other objects in space? _____

   _____

5. The amount of gravitational attraction that exists between objects depends on what two things?_____

   _____

6. What term is used to describe when an object is moving and it keeps on getting faster and faster? _____

7. Is this true or false? Galileo noted that as long as the force of gravity kept being applied to a rolling ball on a ramp, the ball continued to accelerate. _____

8. Is this true or false? Galileo knew that friction and other things can get in the way of how falling objects move and cause them to slow them down or stop. _____

9. Aristotle and Galileo were both scientists, although Aristotle lived hundreds of years before Galileo. Why was Galileo a better scientist that Aristotle?_____

   _____

   _____

   _____

10. Look at the numbers in Table 5.1 showing the distance a ball rolled each second. See if you can complete the line for five seconds on the Table, based on the patterns for the first four seconds. _____

    _____

    _____

☑ **Dig Deeper**

☐ Dig Deeper completed in notebook.

Date:

The Activity:
Procedure and Observations

Part A (Demonstration)

The teacher will tie a short rope to the handle of a bucket of water and hold the rope. The teacher will begin swinging the bucket in a circle going near the floor and then overhead. Observe carefully and see what happens to the water when the bucket is upside-down.

1. When you observed the bucket of water upside-down, did you observe any of the water falling out of the bucket? _____

2. Try to think of a logical reason why it didn't fall out. _____

_____

Part B

Follow the directions for suspending a flexible insulation tube between two chairs so that the ends of the tube are the same height above the floor. Place a marble at the top of one end of the tube. Release the marble. Tell how far it travels up the other side of the tube. Try this several times and alternate sides to release the marble

Part C

Make a ramp from the tube according to the instructions and release the marble from the top. Write your observations about what happens

to the marble when it reaches the end of the tube and moves through the air. Does gravity pull it straight down?

3. Try to make a diagram showing the path of the marble as it reaches the end of the tube and moves through the air until it hits the ground.

Drawing Board:

✏ Stumper's Corner

1. _____

_____

2. _____

_____

1. Is this true or false? Once a ball starts to move, it will continue to move forever unless something causes it to stop. _____

2. What are a few things that could cause a rolling ball to stop moving?

_____

_____

3. All matter has this property: Moving objects tend to stay in motion and objects that are not moving tend to keep on not moving. What is the name of this property? _____

4. Isaac Newton wrote the three laws of motion. What other scientist did research about moving objects that Newton studied and expanded on? _____

5. Aristotle based many of his conclusions about science on logic. On what did Galileo base most of his conclusions about science? _____

_____

6. What did Newton mean when he said, "If I have seen further than others, it is by standing upon the shoulders of giants?" _____

_____

_____

7. Explain why a heavy rock and a light-weight rock will hit the ground at the same time if they are dropped from the same height. _____

_____

_____

_____

8. Complete the path of a moving ball as it leaves the ramp. Draw arrows to show the force of gravity on the ball as it leaves the ramp. Draw arrows to show the effect of inertia as it leaves the ramp.

Drawing Board:

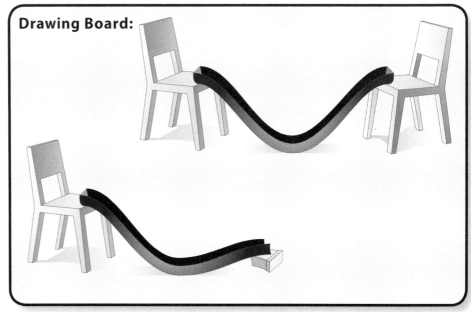

9. In an activity you did, a marble rolled down a piece of tubing. What force caused it to roll down? _____

10. In this same activity, the marble continued to roll and moved up the tubing for a ways. What property of matter caused it to keep rolling up a hill for a ways? _____

11. What effect did the force of friction have on the marble's motion?

_____

_____

_____

Investigation #7
Making Telesopes

Date:

The Activity:
Procedure and Observations

Part A:

Lay a thin convex lens on lined paper and count the number of lines you can see within the lens. How many lines did you see in the lens?

1. _____

Raise the lens up a few centimeters until the lines on the paper are in focus. How many lines did you see this time in the lens?

2. _____

Divide the first number into the second one. This is an estimation of the lens magnification.

3. _____

Now repeat this process for the thick convex lens.

1. _____

2. _____

3. _____

Part B:

Hold one of the magnifying lenses a few centimeters from your eyes. Hold the other magnifying lens at arm's length in front of your eyes. Now look through both lenses at the same time and slowly bring the outermost lens toward your eyes. There should be a point at which you are able to focus on an object in front of you.

1. Does the image appear larger than the object does without the lenses? _____

2. Is the image upside-down or right-side up? _____

If you have a thick lens and a thin lens, repeat this activity, but put the thicker (or smaller) lens near your eye and hold the thinner (or larger) lens at arms' length.

3. Does the image appear larger than the object does without the lenses? _____

4. Is the image upside-down or right-side up? _____

_____

1. _____
   _____

2. _____
   _____

? What Did You Learn

1. Draw diagrams of the two concave lens.

Drawing Board:

2. Which kind of lens is a magnifying glass? _____

3. What is another word for bending light?_____

4. Draw diagrams showing light passing through convex and concave lenses and bending.

Drawing Board:

5. What kind of image would someone see if they used two convex lenses to focus on an object (like in a refracting telescope)?_____

   _____

6. What is one difference in a refracting telescope and a reflecting telescope? _____

   _____

7. What famous scientist built a telescope that enabled him to see some of the moons around Jupiter?_____

8. What is the difference in how a refracting telescope and a microscope are made?_____

   _____

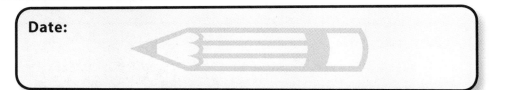

Date:

☐ Dig Deeper completed in notebook.

The Activity: Procedure and Observations

Find the pattern in the Student Journal to make a paper airplane (or use another model if you prefer). Follow the directions to make the paper airplane and try it out. Do several test throws to find the best place to hold the airplane as you throw it. It will probably be near the front end of the plane. When you are ready to test the distance it will travel, use a metric tape measure and record the distance in meters.

Test your glider by throwing it three times and each time measuring the distance it can glide. Record your results on the data table. Place a paper clip in different places on the glider. Follow the directions for testing each of the different places. Compare the results in the data table and try to decide which condition allowed the glider to travel the longest average distance.

	1st Distance	2nd Distance	3rd Distance	Average distance
No Paper Clip				
Clip in Position 1				
Clip in Position 2				
Clip in Position 3				

1. Did any of the paper clips you added help the plane to glide farther? If so, where was the position or positions? _____

_____

2. Did any of the paper clips you added cause the plane to glide a shorter distance? If so, where was the position or positions?

_____

_____

3. Did the plane glide farther with no paper clip? _____

Stumper's Corner

1. _____

_____

2. _____

_____

1. In what year did Orville and Wilbur Wright make their first flight on a glider that was powered by a gasoline motor? _____

2. What did the Wright brothers have to spend months doing before they were able to test their first airplane? _____

_____

3. Gliding was a popular hobby in the 1800s. Explain how a person could use a glider and travel hundreds of feet through the air. _____

_____

_____

_____

4. Who discovered the principle of floating in air? In about what year was this discovery made? _____

_____

5. Another popular hobby that began in the late 1700s was hot-air ballooning. What chemical eventually replaced the hot air and allowed people to travel long distances through the air? _____

_____

6. Commercial air ships were beginning to be used in the early 1900s to carry passengers long distances through the air. What famous air ship burned and crashed, putting an end to travel by air ships? _____

_____

7. Why were people never able to attach artificial wings to their bodies and fly? _____

_____

8. Could an airplane be built that would take a man to the moon? What kind of device was used to transport men to the moon? _____

_____

_____

_____

9. List 3 inventions that had to be developed before someone could build a motor-powered airplane. _____

_____

_____

_____

_____

Investigation #9
Rockets and Space Exploration

Date:

The Activity:
Procedure and Observations

Follow the instructions in your text for making a small rocket.

A

B

C
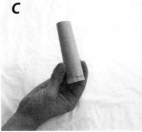

D

E

F

G

H
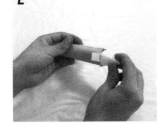

I

Place the rocket on a trash bag on a table. Hold the rocket nose down and pour 1 teaspoon of water into the canister. Now drop in ½ of an effervescent tablet, press on the cap, and set the rocket upright on the trash bag. Record your observations. _____

_____

_____

_____

_____

_____

✏️ Stumper's Corner

1. _____

_____

2. _____

_____

1. State Newton's third law of motion. _____

_____

2. Can airplanes fly through space? Can they fly through air? _____

_____

3. Can rockets fly through space? Can they fly through air? _____

_____

4. In the investigation you did, the contents of the canister pushed down with a certain force. Explain what happened as there was a reaction to this. _____

_____

_____

5. What did Werner Von Braun do during World War II as a rocket scientist in Germany? What did he do in America after the war? _____

_____

_____

_____

6. What projects did NASA accomplish regarding men in space? What is the official name of NASA? _____

_____

_____

7. What happens when a rocket leaves space and begins to re-enter the air around the earth? _____

_____

_____

_____

_____

8. Name a harmful way rockets have been used in history. _____

_____

9. Name two or more useful ways in which rockets have been used in history. _____

_____

_____

_____

_____

Dig Deeper

☐ Dig Deeper completed in notebook.

Date:

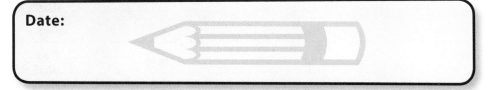

The Activity: Procedure and Observations

Locate the earth's geographic North Pole and the geographic South Pole. The imaginary line running through the center of the earth from the two poles is called the axis of the earth.

Now spin the globe. Notice that the globe is turning on its axis.

Find approximately where you live on earth. Put a sticky note on the globe over your home. Shine a flashlight on the globe and make the globe turn around. Notice when your home is in daylight and when it is in the dark.

At the same time the earth is spinning on its axis, the moon is orbiting the earth, and the earth is orbiting the sun. Try to demonstrate all these motions at the same time, letting the globe represent the earth, the flashlight represent the sun, and a small ball represent the moon. It will take more than one person to demonstrate all of this motion.

The sun was the only thing that was not moving in your demonstration. Do you think the sun actually stands still in space? _____

_____

_____

_____

✎ Stumper's Corner

1. _____

   _____

2. _____

   _____

1. What is the axis of the earth? _____

_____

_____

2. How many times does the earth spin on its axis during a 24-hour day?

_____

_____

3. How long does it take for the moon to make one orbit around the earth? _____

4. How long does it take for the earth to make one orbit around the sun? _____

5. Does the sun stay still or does it spin around the galaxy? _____

_____

6. Briefly explain what gives the earth its seasons. _____

_____

_____

7. During which season are the sun's rays that hit the Northern Hemisphere the most slanted? _____

8. Explain why Venus does not have spring, summer, fall, and winter seasons. _____

_____

9. What chemicals are used by green plants to make food and what chemical is given off as a gas? _____

_____

10. Suppose the earth was too close or too far away from the sun. How might that affect the water cycle? _____

_____

_____

_____

_____

11. List two or three other conditions that are "just right" on earth or in our solar system, such that life would be difficult if they were different. _____

_____

_____

_____

_____

_____

Investigation #11
The Earth's Atmosphere

☐ Dig Deeper completed in notebook.

Date:

The Activity:
Procedure and Observations

Fill the small cup completely full of water and put the index card over the open end of the cup. Hold it over a container to catch the water if there is a spill. Now turn the cup upside-down, holding the card in place with one hand. Don't squeeze the cup as you are holding it. Carefully remove your hand from the card.

What happens to the water in the cup? _____

_____

_____

Try to think of an explanation for why the water didn't fall out of the cup. _____

_____

_____

Stumper's Corner

1. _____

_____

2. _____

_____

1. How do plants and animals maintain a balance of the amount of carbon dioxide and oxygen in the air? _____ _____

2. How much of the air we breathe is made up of oxygen?_____ _____

3. What causes air pressure on the earth? What is the average pressure at sea level? _____ _____

4. Could astronauts survive at the edge of the atmosphere without a pressurized cabin or suit? _____

5. Why do planets and moons with no atmosphere or very thin air not have liquid water on them?_____ _____

6. On the earth, how does the atmosphere affect the temperature at night when there is no sunlight on the earth? _____ _____

7. Briefly explain how the moon's temperature is different from the earth as a result of not having an atmosphere. _____ _____ _____

8. Why does the earth have weather conditions, such as winds, clouds, or rain, while these conditions are not found on the moon?_____ _____ _____

9. Why is the earth not covered with craters like the moon?_____ _____ _____

10. Name the five layers of the atmosphere in order starting with the lowest level. _____ _____

11. Suppose you hiked to the top of a mountain that is two miles high. What differences would you notice in the temperature during your hike? _____ _____ _____

12. Which layer of the atmosphere is composed mainly of small amounts of hydrogen and helium, the lightest gases on earth?_____ _____

13. What happens to charged particles from the sun as they get near the earth? What can often be seen in the sky when this happens? ___ _____ _____ _____

14. From what danger does the earth's magnetic field protect the earth? _____ _____

☑ **Dig Deeper**

☐ Dig Deeper completed in notebook.

Date:

The Activity: Procedure and Observations

Part A

Put a quarter flat on the table to represent earth. Place a penny flat on the table with the "heads" side facing up to represent the moon. Slide the penny around the quarter in a circle, like it was completing an orbit.

Lincoln's face on the penny should keep looking at you throughout its orbit. Make four drawings of the penny and the quarter. Show when the penny is ¼, ½, ¾, and all the way around the quarter.

Repeat this exercise, but this time make sure Lincoln's face on the penny (the moon) always looks at the quarter (the earth) throughout its orbit. Make four drawings of the penny and the quarter. Show when the penny is ¼, ½, ¾, and all the way around the quarter.

Part B

This activity will demonstrate the effects of meteoroids landing on the moon. Take a large flat cookie sheet, fill it almost to the rim with white flour, then cover the surface of the flour with the red paprika. Stand above the cookie sheet, and drop marbles of various sizes onto the surface. Carefully pick up the marbles after they are dropped, trying not to disturb the flour. Observe the craters and rims that form from the impact. Try to have some impacts that overlay previous impacts. Repeat as needed.

Part C

The first part of this investigation will need to be done when there is a full moon that is visible. Measure the diameter of a dime with a metric ruler. Use clear tape to stick the dime to a window where a full moon can be seen. View the moon with one eye so that the dime just blocks out the full moon. Move closer or farther away from the dime until you find this position. Get a partner to help you use a metric tape (not a stiff ruler) to find the distance from beside your eye to the dime. Record this information. Divide the diameter of the coin into the distance from the coin to the eye and record your answer.

Hint: If all your measurements were accurate, you should have found that the diameter of the dime is about 18 mm and your calculated answer is about 110. If you repeat this activity with a larger coin, you will still get the calculated answer of about 110. This will work with a larger coin, because the distance from your eye to the coin will have to increase to just block out the full moon.

1. _____

_____

2. _____

_____

? What Did You Learn

1. Does the moon orbit the earth while the earth orbits the sun? _____

2. How long does it take the earth to make one complete trip around the sun? About how long does it take the moon to make one complete trip around the earth?_____

_____

_____

3. Both the earth and the moon rotate on their axis. How long does it take for the earth to make one complete rotation? About how long does it take for the moon to make one complete rotation? _____

_____

_____

4. Why is only one side of the moon seen from the earth? _____

_____

5. Sometimes only part of the moon is visible in the night sky from the earth. What determines which part of the moon is visible at night?

_____

What Did You Learn ?

6. How does the earth's gravitational pull affect the moon? How does the moon's gravitational pull affect the earth?_____

_____

_____

7. What does it mean to say the angular size of the moon and the angular size of the sun are nearly equal? _____

_____

_____

8. Although there are no lakes or rivers on the moon, is there any indication of water on the moon? Explain._____

_____

_____

9. Circle the things that have not been found on the moon: gravity, air pressure, lakes, wind, craters, mountains, sedimentary rocks, carbon.

_____

_____

10. Equipment left on the moon by astronauts allows scientists to measure the distance from the moon to the earth. Is the moon getting closer to the earth, getting farther away from the earth, or staying the same? _____

_____

_____

☑ **Dig Deeper**

☐ Dig Deeper completed in notebook.

Date:

The Activity: Procedure and Observations

Complete as much of the chart about Mercury, Venus, and Earth as you can. Use the most up-to-date references you can find. You may be able to find the most recent information from websites like <www.NASA.org>.

	Mercury	Venus	Earth	Mars
Magnetic Field				
Composition of Atmosphere				
Length of Day (Compared to Earth)			24 hours	
Length of Year (Compared to Earth)				
Size				
What Would 100 lbs. Weigh			100 lbs.	
Water				
Volcanoes, Earthquakes				
Moons				
Seasons due to a tilt in planet's axis?				
Where Sunset Is Seen				
Landforms and Surface Makeup				
Avg. Surface Temperature				

Stumper's Corner

1. _____

_____

2. _____

_____

? What Did You Learn

1. Which of the rocky planets have the strongest magnetic fields?

_____

2. Why is Venus the hottest planet in our solar system? _____

_____

3. Which of the rocky planets has liquid water on its surface? _____

4. On which of the rocky planets is there a sunset in the east? _____

5. Which of the rocky planets has the longest day? Which one has the shortest day? _____

_____

6. Which of the planets does not have spring, summer, fall, and winter seasons? _____

7. In what year did the Soviet Union first launch a spaceship that flew past the moon? _____

Mars

Earth

Venus

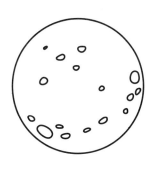

Mercury

8. Do all the rocky planets have an atmosphere? Are they all made up of the same gases? _____

_____

_____

9. What did the MESSENGER satellite discover about the planet Mercury that may indicate it is cooling and shrinking? _____

_____

☑ **Dig Deeper**

☐ Dig Deeper completed in notebook.

Date:

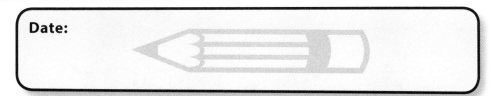

The Activity:
Procedure and Observations

For this lesson, you will need to do some Internet research or find some other reference materials. Try to find at least ten headlines from newspapers or magazines having to do with "extraterrestrial life" or life in space or with finding conditions that would support life. Try to find both old and recent articles. Copy the headlines and any sub-headlines. Give the name of the newspaper or magazine and the date it was published. Write your thoughts about whether the headlines are justified or not.

HEADLINE

Name of Paper: Date:

ILLUSTRATION

Stumper's Corner

1. _____
   _____

2. _____
   _____

? What Did You Learn

1. Before the space program began, why did some scientists believe there were living things on Mars? _____
   _____
   _____

2. Have space satellites and probes found evidence of life on any of the planets? _____

3. Some people believe the earth was visited by aliens from outer space who taught ancient civilizations their advanced technologies. What is another explanation for this? _____
   _____
   _____

4. Is one of the goals of the space program to find more about the origin and evolution of everything? _____

5. Do many scientists reject a supernatural explanation for how everything began because they believe there must be a natural explanation for this? _____

6. Briefly tell what happened during the Enlightenment Period of history. _____
   _____
   _____

7. Give one thing proposed by naturalism. _____
   _____
   _____

8. What kind of research does SETI do? _____
   _____

Investigation #15
The Jovian Planets: Jupiter, Saturn, Uranus, and Neptune

☑ **Dig Deeper**

☐ Dig Deeper completed in notebook.

Date:

The Activity:
Procedure and Observations

Complete as much of the chart about Jupiter, Saturn, Uranus, and Neptune as you can. Use the most up-to-date references you can find. You may be able to find the most recent information from websites like www.NASA.gov.

	Jupiter	Saturn	Uranus	Neptune
Magnetic Field				
Chemicals in Atmosphere				
Length of Day				
Length of Year				
Size around Equator				
What Would 100 lbs. Weigh				
Moons				
Characteristics of Moons*				
Rings Around Planets				
Tilt of Axis				
Retrograde Rotation or Like Earth				
Surface Temperature**				
Weather Conditions				

1. _____

   _____

2. _____

   _____

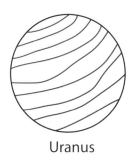

Jupiter   Saturn

Uranus   Neptune

? What Did You Learn

1. Which of the Jovian planets have a strong magnetic field?_____

   _____

2. Which is the largest Jovian planet?_____

3. Which Jovian planet has the most moons? _____

4. Which Jovian planet has the longest day? Which one has the shortest day? _____

5. Which Jovian planets have rings around them?_____

6. Which Jovian planet is most tilted on its axis? _____

7. Which Jovian planet spins on its axis opposite to the way the earth spins? _____

8. Jovian planets are made up of what two main elements?

   _____

9. Where are there huge cyclone storms on Jovian planets?_____

   _____

   _____

10. What is one thing that is unusual about Jupiter's moon Europa?

    _____

    _____

11. What is one thing that is unusual about Neptune's moon Tritan?

    _____

    _____

12. Name two other moons that orbit Jovian planets and give some facts about them._____

Investigation #16
The Sun and Its Light

Date:

The Activity:
Procedure and Observations

Part A.

Read the label on different bottles of sunscreen. Make a chart (or use the one below) comparing the different bottles. Include brand name, SPF rating, the ingredients, and any other features mentioned.

Sunscreen Brand Name	SPF rating	ingredients	other

Part B.

Use the sun, a projector, or a strong flashlight as a source of light. Shine the light on something white. Place the prism in the path of the beam of light and turn it slightly back and forth. Carefully observe the colors that form on the white surface. What do you observe? Make a sketch on your paper of the bands of color you saw. Name each color and show the order in which they appear on the paper. If you have trouble seeing the spectrum clearly, hold the prism next to your eye and look at a light bulb (that is on), and name the colors you see from the light. (Do NOT look at the sun!)

_____

_____

_____

_____

_____

1. _____
   _____

2. _____
   _____

? What Did You Learn

1. Electromagnetic waves travel through space at what speed?
   _____
   _____

2. What are three ways in which electromagnetic waves can differ from each other? _____
   _____

3. Name three kinds of electromagnetic waves that are longer than visible light._____
   _____
   _____

4. Name three kinds of electromagnetic waves that are shorter than visible light._____
   _____

5. Which are more dangerous — very short electromagnetic waves or very long electromagnetic waves? _____

6. Do very short electromagnetic waves have high frequencies (vibrate faster) or low frequencies (vibrate slower)? _____
   _____

7. What kind of electromagnetic waves are filtered out by sunscreen lotion? _____
   _____

8. How do sound waves differ from light waves and other kinds of electromagnetic waves? _____
   _____
   _____
   _____

9. What kind of wave is detected by night-vision goggles?_____
   _____

10. What instrument do astronomers use to produce a spectrum of visible light from the sun and other stars? What information about these stars can they obtain from studying these spectra?
   _____
   _____
   _____
   _____

The Sun and the Earth Relationship

☑ **Dig Deeper**

☐ Dig Deeper completed in notebook.

Date:

The Activity: Procedure and Observations

Part A.

Place a pole or stick into the ground so that it is vertical with nearby buildings. Do this on a sunny day so that the pole's shadow can be seen. Keep the information you collect in a data table. First, measure the length of the pole. Next measure the length of the pole's shadow and record, along with the time and date. Wait 15 minutes and measure again. Take measurements every 15 minutes for at least an hour. Go longer if you can. How long was the shadow when you began measuring it? How long was the shadow after one hour? How much did the length of the shadow change during the time you measured it? If you measured for longer than an hour, what information did you record?

Part B.

An eclipse of the sun occurs when the moon is between the earth and the sun and the moon's shadow falls on the earth. Design and make a model that demonstrates an eclipse of the sun. Explain your model to others.

✏ Stumper's Corner

1. _____

2. _____

1. In what city is there a special telescope that is used to pinpoint noon when the sun is exactly overhead, and is used to set all clocks on earth? _____

2. Before mechanical clocks were invented, what was used as an accurate kind of clock to tell time? _____

3. Which object in our solar system makes up over 99.5 percent of the mass of the solar system? _____

4. What do scientists believe is the source of the energy that comes from the sun? _____

   _____

5. Which kind of electromagnetic waves leave the sun and reach the earth? _____

   _____

   _____

6. When conditions are just right, the moon passes between the earth and the sun and is the exact size to briefly block the sun's light from reaching a small portion of the earth. What is this event called?

   _____

7. What causes an eclipse of the moon to occur? _____

   _____

8. What is a variable star? Is our sun a variable star or a stable star?

   _____

   _____

   _____

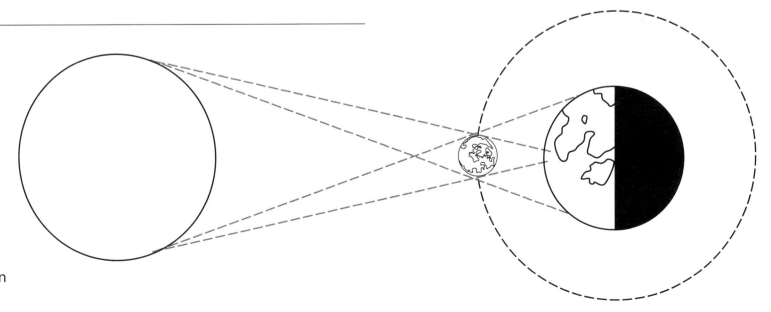

Eclipse of the Sun: Label as in text book.

☐ Dig Deeper completed
in notebook.

Date:

The Activity:
Procedure and Observations

1) Take a large sturdy shoebox with a removable lid (see photos in text book). Cut a small hole in one end of the shoebox. This will be the eye portal that you will look through. On the other end of the box, cut out almost the entire end, but leave a 2 cm border on three sides.

Now cut a piece of cardboard or heavy paper that is the same size and shape as the end of the box. It should be cut out like the end of the box with a 2 cm border on 3 sides. Place it about 4 mm from the end of the box. It should fit snugly in place and be reinforced with tape. This will make a pocket to hold your constellation patterns. Spray paint the inside of your shoebox with a flat black paint to make it nice and dark inside.

2) Measure the size of the opening in the end of the shoebox. Make a template from a piece of cardboard that is about 3 to 4 cm wider than the opening and as tall as the box. Make adjustments until you get a square that is the right size to fit into the pocket. When your template is the right size for the pocket, then cut several squares from the black paper using the cardboard template as a pattern.

Now trace the diagrams of the constellations onto the black squares using a silver marker (see appendix). If you want to find and make other examples of constellations, make sure they fit within the size of your opening. Draw an arrow in the upper right-hand corner

pointing upward. This will always give you the proper orientation for the constellation. Draw all your constellations and label them with a silver marker. Punch holes in the paper with a pencil to represent the stars in the constellation, making some holes smaller and some larger according to the way the real constellation looks in the sky. Some major constellations are: Andromeda, Aquarius, Aries, Cancer, Capricornus, Cassiopeia, Cepheus, Cetus, Corona Borealis, Cygnus, Draco, Eridanus, Gemini, Hercules, Hydra, Leo, Libra, Lyra, Orion, Perseus, Pisces, Sagittarius, Scorpius, Taurus, Ursa Major (Big Dipper), Ursa Minor (Little Dipper), and Virgo. There are 88 standard constellations.

3) When viewing a constellation, remove the lid, slide one of your constellation patterns on the black paper squares into the slot. Replace the lid on the box. Hold the box up to the light and look through the viewing portal. This will give you a good representation of what the constellation looks like in the night sky.

4) Paint and decorate the outside of your shoebox any way you like.

Name the constellations you were able to view in your box.

1. _____

_____

2. _____

_____

1. How many constellations are recognized today by the International Astronomical Union? _____

2. How do scientists and astronomers use the constellations today?

_____

_____

3. How many zodiac constellations are there? Do they appear at predictable times of the year? _____

_____

4. Which is the best explanation for why a different zodiac constellation appears each month — (a) because the stars move around the earth, or (b) because the earth moves around the sun? _____

5. Are the zodiac constellations seen high in the sky or near the horizon?

_____

6. Do the planets always appear in the same constellations or do they appear to wander about in the sky? _____

_____

7. Name three constellations or star clusters mentioned in the Book of Job. _____

_____

8. Explain how to locate the North Star in the sky._____

_____

_____

_____

Date:

The Activity:
Procedure and Observations

Make a collection of pictures of things that have been photographed in space. Try to find at least 15 of the following: nebula, spiral galaxy, elliptical galaxy, ring galaxy, constellation, binary star system, each of the planets (count each planet as a separate picture), the earth's moon, some of the moons of the other planets, our sun, a comet, a red star, a white dwarf star, and a blue star. The photographs can be made in regular visible light or they can be enhanced with ultraviolet or infrared light, or even with x-rays or radio waves. Avoid pictures featured on websites that are not photographs of real space objects, such as animations or artist renderings of things that cannot or have not been observed. Read the captions carefully to make sure they are actual photographs. Label each drawing and briefly give some information about each, including their size.

Choose your top three favorites and give five to ten facts (not hypotheses) about each of them.

Stumper's Corner

1. _____

_____

2. _____

_____

1. Name three natural objects that have been found in space other than stars, planets, and moons. _____

_____

_____

2. Stars differ from one another in many ways. List at least five ways one star might differ from another star._____

_____

_____

3. What are some ways space is being studied and explained today that were not available in the days of Galileo?_____

_____

_____

4. Do creation scientists and evolutionary scientists examine the same facts and observations but come to different conclusions when trying to explain what happened in the past? _____

_____

_____

5. Briefly tell the two main explanations for how things in the universe came to exist. _____

_____

_____

_____

6. Evolutionary astronomers believe stars have a life cycle they go through. Do they believe the life cycle of a star would be thousands or years or billions of years? _____

_____

_____

7. What are some of the ways in which stars may change as they age and get older? _____

_____

_____

8. What do scientists call a star that explodes? _____

_____

_____

_____

9. Compare the temperature of a yellow star, a red star, and a blue star. Our own yellow sun is a medium-hot temperature. _____

_____

_____

_____

Investigation #20
Chaos or Creator?

Date:

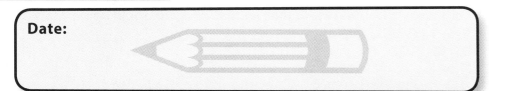

The Activity: Procedure and Observations

Select two or more of the following points and find additional information about them. Write about what you find in your own words.

1. Evolutionary astronomers assume that the more impact craters there are on a moon or planet, the older the object is. Impact craters are found throughout the solar system on planets and their moons, but they are not found evenly all around. Rather than being impacted randomly for billions of years, they seem to have been exposed to a few bursts of meteor showers.

2. Mercury is showing evidence of cracking and shrinking as it cools with vertical cliffs hundreds of miles long.

3. The earth's spin rate is slowing down. In 2008, another leap second was added to the year to take this into account. Secular scientists are puzzled about how fast it might have been spinning billions of years ago.

Stumper's Corner

1. _____

_____

2. _____

_____

1. How does creationism differ from naturalism as a way to explain the beginning of the universe, the world, and living things?

_____

_____

_____

2. What did Dr. Von Braun say about overlooking the possibility that the universe was planned by God? _____

_____

_____

_____

3. Briefly explain three observations that support a young, created universe. _____

_____

_____

_____

Connect the dots and name the constellation. _____

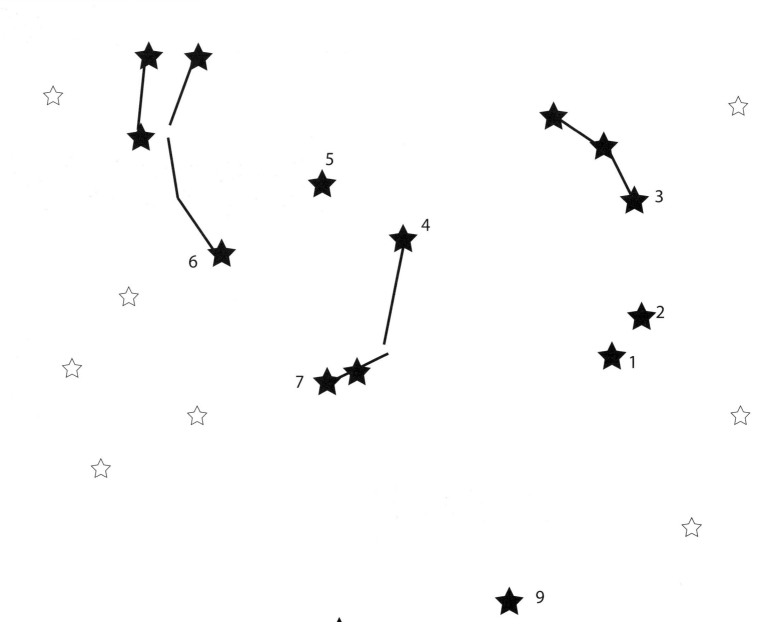

Connect the dots and name the constellation. _____

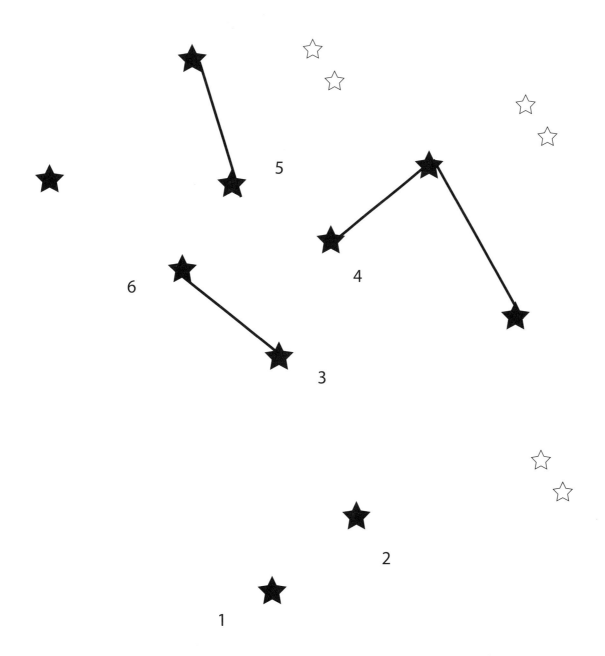

Dashed line indicates fold.

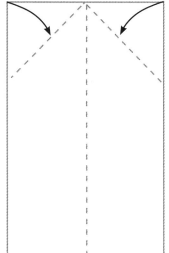

Fold top 2 corners to center. This will make a triangle on top

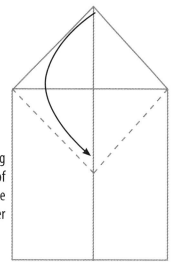

Now fold along bottom side of △ flaps in to the lower paper

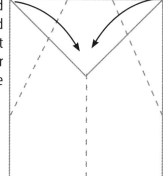

Tricky Part: Fold top corners inward to a point about the middle of your triangle

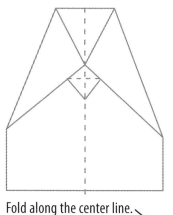

Now fold the excess point of your original △ upward to lock in your previous folds.

Fold along the center line.

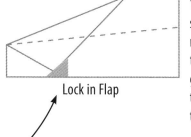

Lock in Flap

Your wing fold should not be par-rallel to the base of the airplane. It will gradually increase toward the back of the body

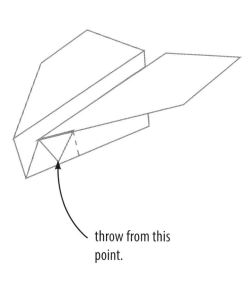

throw from this point.